Contents

What are grasshoppers?

Grasshoppers are **insects**. Insects are small creatures, like flies and wasps, which have six legs. All grasshoppers look similar but there are many different types in the world.

Grasshopper

Karen Hartley,
Chris Macro
and Philip Taylor

First published in Great Britain by Heinemann Library
Halley Court, Jordan Hill, Oxford OX2 8EJ
a division of Reed Educational and Professional Publishing Ltd.
Heinemann is a registered trademark of Reed Educational & Professional Publishing Limited.

OXFORD MELBOURNE AUCKLAND
JOHANNESBURG BLANTYRE GABORONE
IBADAN PORTSMOUTH NH CHICAGO

Designed by Celia Floyd
Illustrations by Alan Fraser [Pennant Illustration]
Printed and bound in Hong Kong/China by South China Printing Co. Ltd.

03 02 01 00
10 9 8 7 6 5 4 3 2 1   500 654076

ISBN 0 431 01698 4
This title is also available in a hardback library edition (ISBN 0 431 01690 9)

British Library Cataloguing in Publication Data

Hartley, Karen
    Grasshopper. - (Bug books)
    1.Grasshoppers - Juvenile literature
    I.Title II.Macro, Chris
    595.7'26

Acknowledgements
The Publishers would like to thank the following for permission to reproduce photographs:
Ardea: p8, J Daniels p17, P Goetgheluck pp5, 11, 13, 14, 26, J Mason p10; Bruce Coleman Limited: J Burton p25, W Cheng Ward p12, M Fogden p7, H Reinhard p18, K Taylor pp6, 24; Garden and Wildlife Matters: p27, S Apps p15, K Gibson p19; Trevor Clifford: pp28, 29; NHPA: S Dalton pp20, 21, 22, H and V Ingen p16; Okapia: P Clay p23, M Wendler p4; Oxford Scientific Films: L Crowhurst p9.

Cover photographs: Gareth Boden (child); J Brackenbury, Bruce Coleman Ltd (grasshopper).

Every effort has been made to contact copyright holders of any material reproduced in this book. Any omissions will be rectified in subsequent printings if notice is given to the Publisher.

Any words appearing in the text in bold, **like this**, are explained in the Glossary.

Some grasshoppers are called **crickets**. Many large grasshoppers are known as **locusts**.

What do grasshoppers look like?

Grasshoppers have long bodies, four long, thin wings and large eyes. They have very hard skin which is green or brown. They have two **feelers** on their heads.

The two back legs of a grasshopper are much longer than the other four. Some grasshoppers look the same colour as the leaves of the plants they live on.

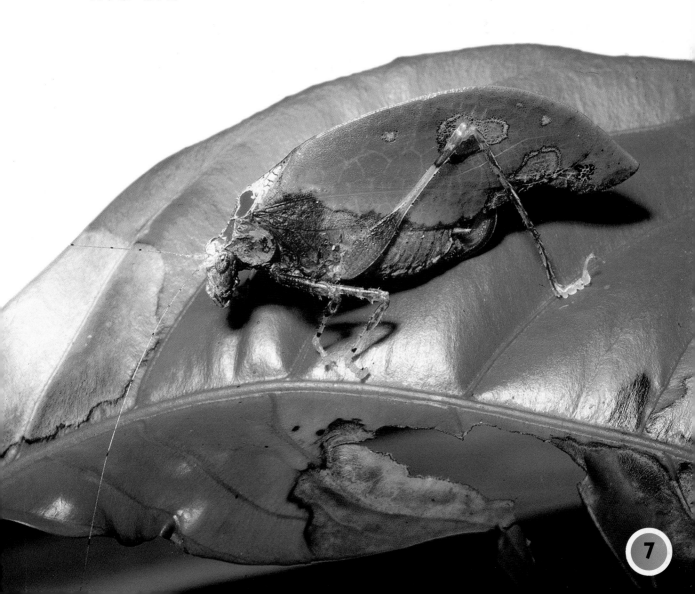

How big are grasshoppers?

Some types of grasshopper are much bigger than others. Many grasshoppers are about as long as your little finger.

The **females** are longer than the **males** by about the length of your little fingernail. Apart from the difference in size, the males and females look alike.

How are grasshoppers born?

The **female** grasshopper lays about a hundred eggs in late summer. She covers them with a sticky liquid. This goes hard and protects them in a **pod**.

The eggs stay in their pod all winter. Young grasshoppers **hatch** out of the eggs in spring. They are very small and look like **adults** without proper wings.

How do grasshoppers grow?

Grasshoppers grow so quickly that their skin bursts and they **moult**. Underneath is a new skin so the grasshopper can carry on growing.

Grasshoppers moult and change their skins four to six times before they are grown up. When they are bigger the new skins have wings attached.

What do grasshoppers eat?

Grasshoppers have strong jaws called **mandibles**. They use them to cut and chew the grass and leaves that they eat.

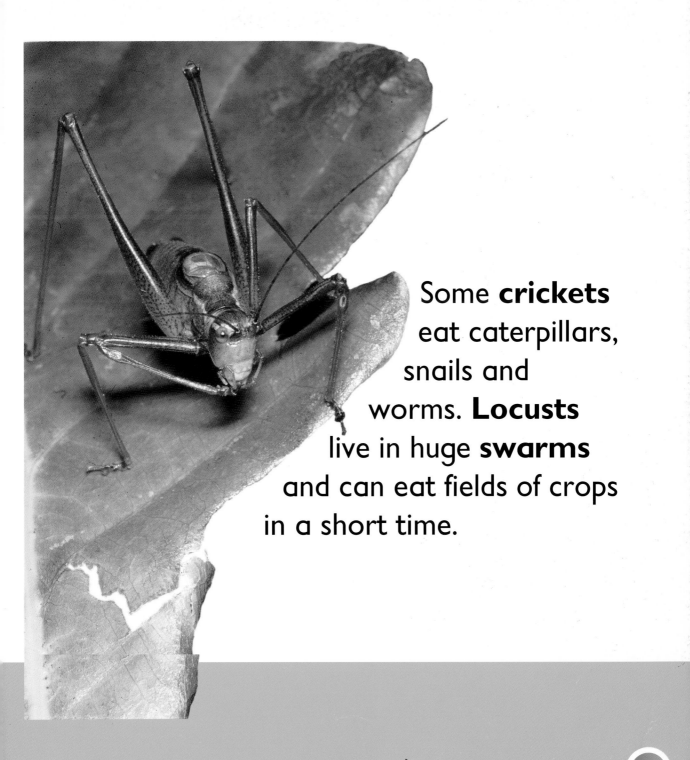

Some **crickets**
eat caterpillars,
snails and
worms. **Locusts**
live in huge **swarms**
and can eat fields of crops
in a short time.

Which animals attack grasshoppers?

Spiders eat grasshoppers if they land on their webs. Frogs and newts eat grasshoppers. They catch them with their long tongues. Birds, snakes and lizards will also eat grasshoppers.

Most grasshoppers are attacked when they are still in the egg, or when they are very young. Grasshoppers lay large numbers of eggs so that some will survive.

Where do grasshoppers live?

Grasshoppers live in nearly all countries. Most grasshoppers live in thick grassland or woodland. Other types live in sand dunes and some live on cliffs.

Some grasshoppers live in houses and a few types live under the ground. Other grasshoppers live in special places like the banks of streams.

How do grasshoppers move?

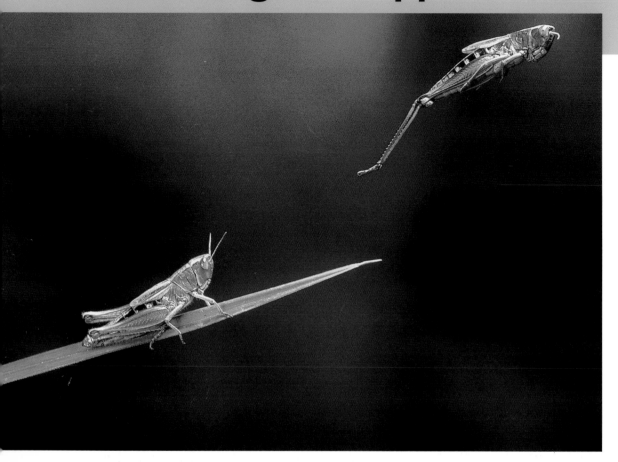

Grasshoppers usually move by jumping. They have very long back legs to give a strong push off the ground. They jump so high it is like an **adult** person jumping over a house.

Some adult grasshoppers have wings as long as their bodies. Although they have large wings, only **locusts** are good flyers.

How long do grasshoppers live?

Soon after the **female** grasshoppers have laid all their eggs, both the **males** and the females die. Sometimes female **crickets** eat the male crickets before they die themselves.

Grasshoppers cannot live through a cold winter but the eggs survive and **hatch** out in the spring. Most of these young grasshoppers will die later in the year.

What do grasshoppers do?

When grasshoppers are fully grown they spend a lot of time on grass stalks. The **males** make a noise called 'singing' to attract the **females.**

The singing can last all day and into the night. Grasshoppers make more noise if it is sunny and dry. If it is raining and cold they are quiet.

How are grasshoppers special?

When **male** grasshoppers sing, they rub the bumps on their back legs along the hard edges of their wings to make loud sounds.

The song is different for each type of grasshopper and the **females** know which is the right song for them.

Thinking about grasshoppers

These two children are looking for
grasshoppers they can hear in the
grass. They want to watch what the
grasshoppers do for a day or two.

They have brought a plastic tank with them. What would they need to put into the tank so the grasshoppers could live there?

Where would be the best place to put the tank?

Bug map

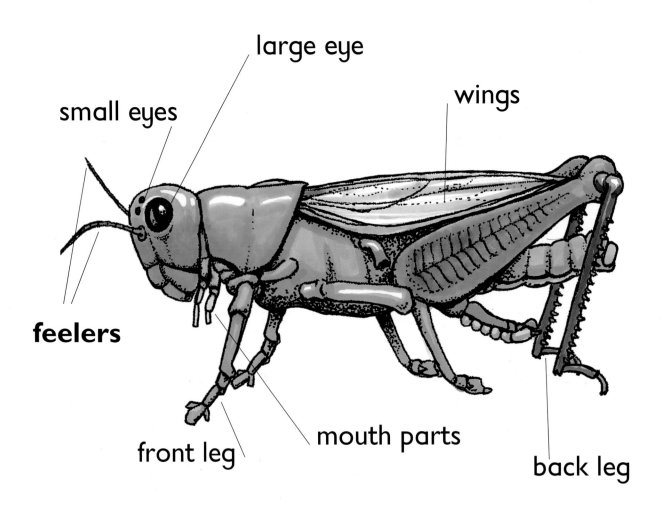

small eyes

large eye

wings

feelers

front leg

mouth parts

back leg

Glossary

adult a grown-up

cricket a type of grasshopper with very long feelers

feelers thin growths from the head of an insect that help the insect to know what is around it

female a girl

hatch when an animal comes out of its egg

insect a small animal with six legs

locust types of large grasshoppers, usually living in hot countries

male a boy

mandibles parts of the mouths of insects, used for biting and chewing

moult when a grasshopper grows too big for its skin, it grows a new one and slides out of the old one

pod a case around eggs or seeds, which helps to keep them safe

swarm many, many insects, or other animals, which fly around together in one large group

Index